Who Am I?
Swamp Animals

Written by Read With You
Center for Language Research and Development

The characters and events portrayed in this book are fictitious. Any similarity to real persons, living or dead, is coincidental and not intended by the author.

Published by Read With You Publishing

Designed by Read With You Center for Language Research and Development

ISBN-13: 978-1-944710-82-8
ISBN-10: 1-944710-82-5

Printed in the United States of America.

I have bumpy skin.

Who am I?

toad

I walk on water.

Who am I?

water strider

I smell with my tongue.

Who am I?

lizard

I have a powerful jaw.

Who am I?

snapping turtle

I’m also called a lightning bug.

Who am I?

firefly

I can swim backwards.

Who am I?

crayfish

I don’t have ears.

Who am I?

snail

Only the female drinks blood.

Who am I?

mosquito

Toad

Water Strider

Lizard

Snapping Turtle

I am a toad.

I am a water strider.

I am a lizard.

I am a snapping turtle.

Firefly

Crayfish

Snail

Mosquito

I am a firefly.

I am a crayfish.

I am a snail.

I am a mosquito.

We are swamp animals.

A swamp is a wetland with trees and plants.

Made in United States
Orlando, FL
07 November 2026

90337523R00017